glyn valley

coaches

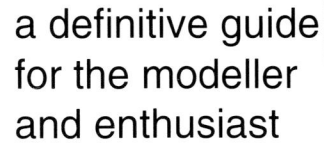

a definitive guide
for the modeller
and enthusiast

by
Mike Higgins & Bernard Rockett

Theodore Press, Bridgnorth, Shropshire

Copyright © Mike Higgins and Bernard Rockett, 1995

All Rights Reserved. No part of this publication may be reproduced, stored in a retrieval system, or transmitted in any form or by any means, electronic, mechanical, photocopying, recording or otherwise, without prior written permission of the copyright holders.

First published in Great Britain in 1995 by Theodore Press
British Library Catalogue record for this book is available from the British Library.

ISBN 0-9523223-0-7

Design and drawing by Theodore Press

Printed and bound in Great Britain by Printpoint, Bridgnorth, Shropshire.

Theodore Press
Orchard House, Manor Farm Lane, Oldbury, Bridgnorth, Shropshire. WV16 5HG

CONTENTS

- 3 Acknowledgements
- 4 Introduction
- 5 Horse-drawn Coaches
- 8 Carriages of the Steam Era, The Clerestory Saloon
- 11 Centre-door Closed Saloon
- 15 Small Open Carriage with End Balcony
- 16 1892 Series Third Class Compartment Carriages
- 19 Open Third Class Carriages
- 23 The Closed First Class Carriage
- 25 1901 Series Third Class Compartment Carriages
- 30 Coach Livery
- 39 Brakes, Running Gear and Couplings
- 45 Closure and Dispersal of the Stock
- 48 Restoration and Preservation on the Talyllyn Railway
- 49 The Glyn Valley Tramway Group
- 50 Coach Numbering
- 51 Modelling the Glyn Valley Tramway Coaches

Acknowledgements

The authors wish to offer their sincere thanks to the Glyn Valley Tramway Group, the Merioneth Railway Society and to the following individuals for their assistance and permission to use unpublished information; Eric Lloyd, David Ll Davies, David Newham, David Bovington, Alistair Parsons, John Stitson, Trevor Cousens. Our grateful thanks are due to Val Rockett for preparation and typing of the manuscript.

Introduction

The Glyn Valley Tramway closed over fifty years ago and while the story of its history and operation has been told in several places, memories of the rolling stock are growing dim. This account brings together much of the available evidence on the coaches that were used on the Tramway. Information has been brought together from contemporary documents and photographs, from those who worked on the railway or who knew it as passengers and from those who observed the gradual disintegration of stock in gardens or farmyards after closure.

All of the coaches built for the Tramway were of four-wheel design, no six-wheeled or bogie coaches were used even though these were often employed on other narrow gauge lines in Wales. The coaches, once delivered to the line, were not turned and footsteps were fitted on only one side, the north or "valley" side. Entry and exit thus took place on this side when no platform was available to passengers. At Chirk station, the valley side of the coach was the side opposite to the platform. These arrangements were made in order to comply with the Board of Trade Regulations which stated that *"every carriage used on the tramways shall be constructed so as to provide for the safety of passengers and their safe entrance to and exit from."*

The authors hope that readers with new information, documents or photographs or with corrections to any errors in the text will feel free to contact them. The reference numbers given to photographs are taken from the Glyn Valley Tramway Group Photograph List.

Horse-drawn coaches

The construction of the Glyn Valley Tramway from Gledrid near Chirk to the Cambrian Slate Quarries at Glyn Ceiriog was largely complete by the spring of 1873 and the carriage of goods traffic began immediately. Just a year later, a minute in the Companies records reported: *"at the request of the inhabitants at Glyn a passenger car had commenced running between Pontfaen and the New Inn, and that Mr Jones had ordered a suitable open car from Messrs Ashburys"*.

Initially, passengers were conveyed in slate wagons because the Company had no suitable vehicles and the first coach acquired from Ashburys in 1874 was something of a stop-gap. It was an open four-wheeled vehicle with a canvas roof supported by iron stanchions. In wet weather canvas side screens could be unrolled and lowered to protect passengers. The driver was placed on a high seat at the front in the style typical of a road coach of the period, indeed the vehicle would have appeared much more at home on the road than on the railway.

"Ashburys" were the *Ashbury Railway Carriage and Wagon Co. Ltd.*, of Openshaw in Manchester, a large, active organisation which supplied a number of coaches to the Tramway. Messrs Ashburys were acquired by the Metro-Cammel company in 1902, the Openshaw works eventually closed in 1925. The open, or toast-rack, coach provided a hasty interim solution to the Board of Trade requirements for the provision of a purpose-built passenger vehicle. Few photographs of it have been traced, in the earliest, (GVTG No. 56), the canvas roof with rolled side screens are in place and a small boy poses proudly with the driver, a single horse was used to provide the motive power.

Two small closed coaches were supplied by Messrs Ashburys a short time later. These had the appearance and form of railway carriages, the sides were plain with rather small windows placed high up and near to each end as well as in the middle, the two separate compartments were each accessed by an unglazed door. The ends were panelled and illumination for the passengers was provided by a single, centrally mounted, oil lamp.

When the Tramway was converted to steam traction in 1891, the three horse-drawn coaches were adapted for their new role. A photograph of March 1891 (GVTG No. 109) shows the inaugural steam train with a closed coach in the formation ahead of the toast-rack coach, both looked completely out of place coupled between the large heavy coaches of the steam era.

The annual returns of the Company suggest that the last horse tramway coach had been retired from active service by 1894 but they continued to be mentioned in records until 1923. A photograph taken in 1929 of a train leaving Chirk (GVTG No.52) shows the grounded body of a closed coach on the embankment south of the road bridge in use as a platelayers hut. The second closed coach was reputed to have ended its days at Castle Mill in the same way.

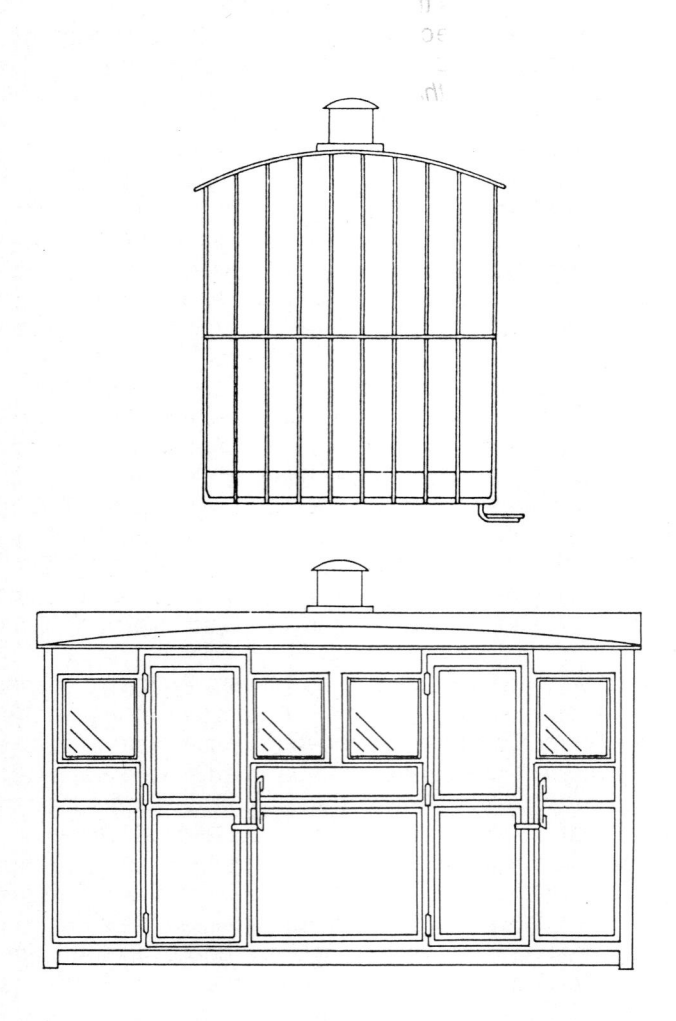

ca. 1881 Horse tramway closed two compartment coach, underframe details are not known.
Scale 7mm = 1ft

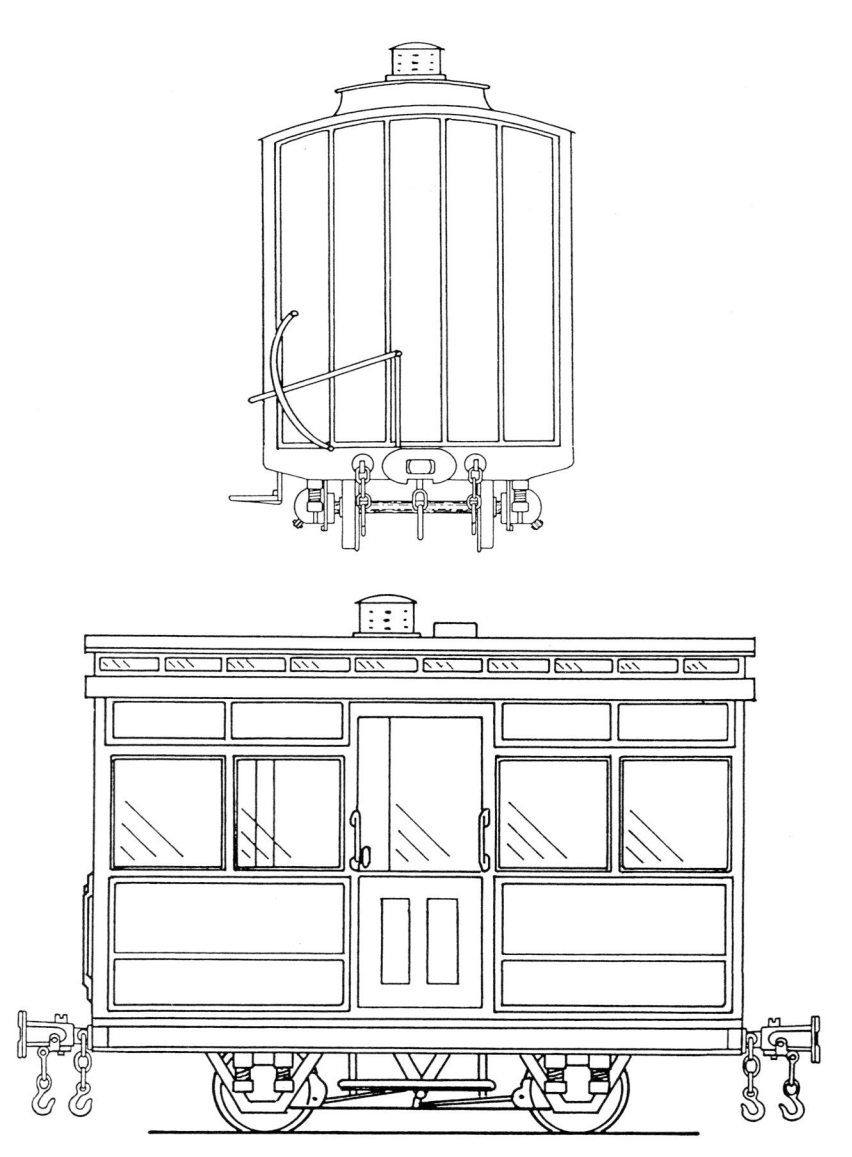

1891 Closed clerestory saloon in later rebuilt form with centre oil lamp.
Scale 7mm = 1ft

Carriages of the steam era

Over a ten year period from 1891, fourteen coaches were taken into service on the Tramway, these included six open carriages, four closed saloons and three assorted vehicles all of third class. A single first class saloon completed the stock. Each type will be described in chronological order.

The clerestory saloon

The clerestory coach was one of three non-standard carriages acquired by the GVT in 1891. A quotation was received in 1889 from the *Midland Carriage and Wagon Company* of Shrewsbury for the supply of two carriages at a cost of £150 but the Company appears to have delayed placing a definite order until 1890. It seems that it was hoping to use the old horse-drawn stock for the first few years of steam operation but was pressed by the Board of Trade to obtain suitable carriages for use with steam locomotives. With the opening of the new service in March 1891, the GVT found it difficult to obtain stock in the short time available and was obliged to take what was available, a clerestory saloon, a centre door saloon and an end balcony open carriage.

As delivered, the clerestory saloon was a fine example of the coach builders craft, it was fully glazed and generously lined out in gold leaf following the best tramway tradition of the late nineteenth century. It carried a full height sliding door in the centre of each side which was glazed to match the four windows. The shallow clerestory roof had ten small glazed panels on each side, some of these could be opened for ventilation. Slatted wooden seats were arranged lengthwise inside the vehicle, passengers sat back-to-back. A gangway separating the seats into two groups was left between the doors. The interior was finished in varnished wood with horizontal boards at each end and

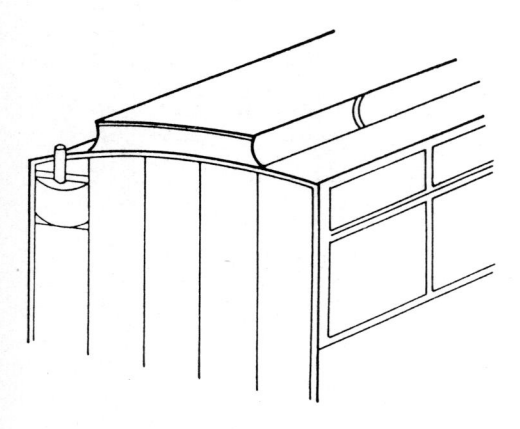

Detail of original end lamp mounting on 1891 clerestory coach. A lamp was fitted at each end

illuminated by oil burning coach lamps placed high on the left-hand side of each end. The lamps were accessible from outside the vehicle.

In service, the clerestory saloon posed a number of problems, ventilation was poor, the sliding door was considered unsafe and access for passengers and maintenance staff was difficult. These caused a programme of modifications to be carried out around the turn of the century. The clerestory roof was reconstructed in a narrower, higher form and a single Colza oil lamp was fitted in the centre of the roof to replace the two end lamps. Steps and a handrail were added to the end of the carriage to allow the lamplighter to gain access to the suitably strengthened roof, also a passenger footstep was fitted below the door on the "valley side".

A few years later, but before 1905, the sliding doors were cut in half at the waistline, the lower parts were hinged and refitted in stable door fashion while the upper parts remained as sliding panels to provide ventilation. Droplights would have been more conventional but this compromise was certainly cheaper. An interesting woodcut appears in the *W.G. Bagnall Ltd.* catalogue for 1905 in which the clerestory saloon is depicted in "a train of mixed traffic for 2'6" gauge" and is labelled "workman's carriage". All of the modifications mentioned above are clearly visible. A similar illustration was used in the magazine *"Machinery Market"* of 1895 in an article on the Bagnall company. It seems likely that as well as acting as selling agents for the *Midland Railway Carriage and Wagon Co. Ltd.* at this time, Bagnalls may have supplied some of the running gear for the coach.

Photographs show that the vehicle was in regular use right up to the withdrawal of passenger services in 1933 and a late picture (GVTG No. 28) shows it awaiting disposal at Chirk in 1936.

Dimensions of the Clerestory Saloon

Length	- overall	14ft	3in	Height - rail to roof		8ft	5in
	- over frames	11	7	- rail to coupling centre		1	11½
Width	- over mouldings	5	10	Wheelbase		5	6
				Wheel diameter		1	9

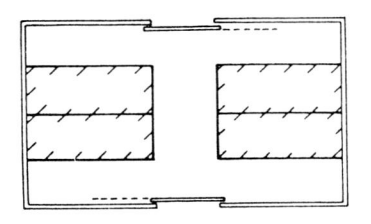

Seating arrangement in clerestory saloon for 12 passengers

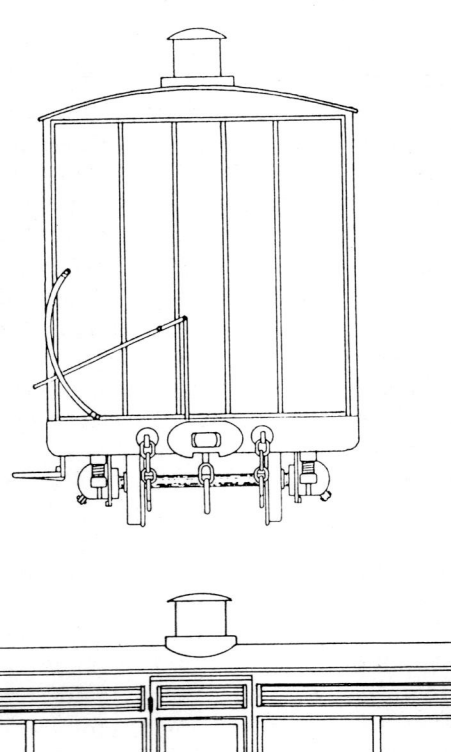

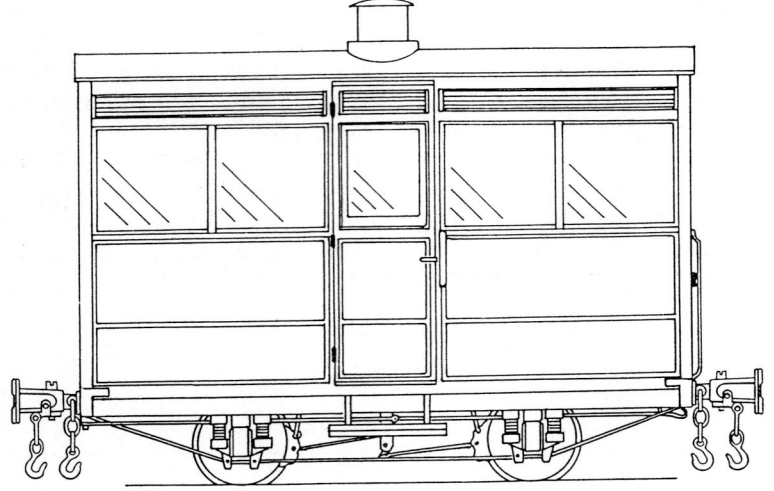

1891 Closed centre door saloon, later designated third class.
Scale 7mm = 1ft

Centre-door closed saloon

This coach was the second of the three odd coaches purchased by the GVT at the beginning of steam haulage. It was delivered somewhat later than the clerestory saloon and since it was built by the same company, several of the modifications made later to this coach were incorporated into the new one. The overall dimensions were similar, the body was constructed on the same pattern of underframe, the centre door was hinged conventionally at the side and carried a full-width glazed droplight.
The roof was plain with a centrally mounted Colza oil lamp, the coach sides were panelled below the waist with glazed lights above. Just below the roof a louvred ventilator extended across the full width of each side including the door. Slatted wooden seats were arranged lengthwise inside the coach and the other interior fittings were also like those used in the clerestory coach. A passenger footstep was installed on the valley side only and access to the roof was by footsteps and a handrail on one end.

Dimensions of the Centre-door Closed Saloon

These are identical to those of the clerestory saloon although the roof does not, of course, carry a clerestory structure.

Glyn Valley Tramway closed coaches were built with ends that extended below the corresponding sides. Matching of sides and ends at the four corners was achieved by tapering the bottom of each end. However, this closed coach was an exception to the rule in having rounded ends at the bottom. The end steps provided another peculiarity in that each was bolted to the coach body above the step. All other coaches fitted with roof lamps carried end steps that were bolted below the step. The vehicle enjoyed a long life and remained in service alongside its twin the clerestory coach for some forty five years until the closure of the line.

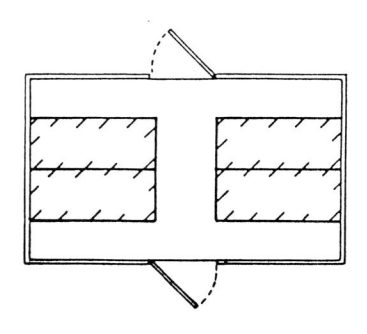

Centre door saloon seating 12 passengers

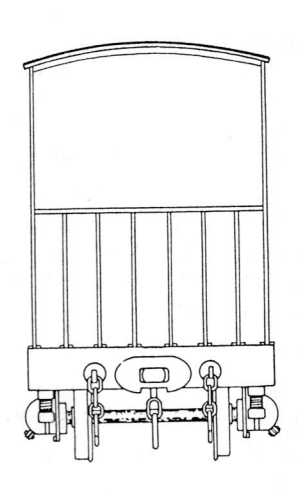

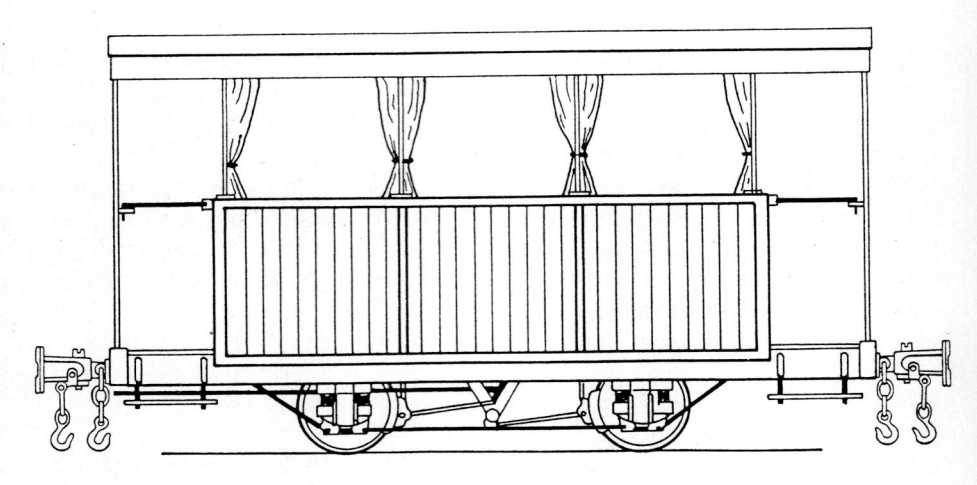

1891 Small end balcony open saloon, early form with canvas side curtains and open ends.
Scale 7mm = 1ft

Nine coaches stand in the loop outside Chirk Station after closure of the tramway, 1935 (Glyn Valley Tramway Group).

Five closed coaches stand at Chirk Station, the 'Garter and Crest' livery is just visible, ca.1929 (MJ Higgins collection)

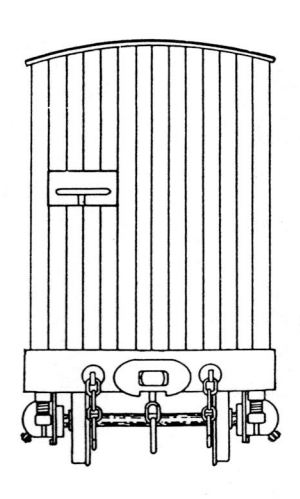

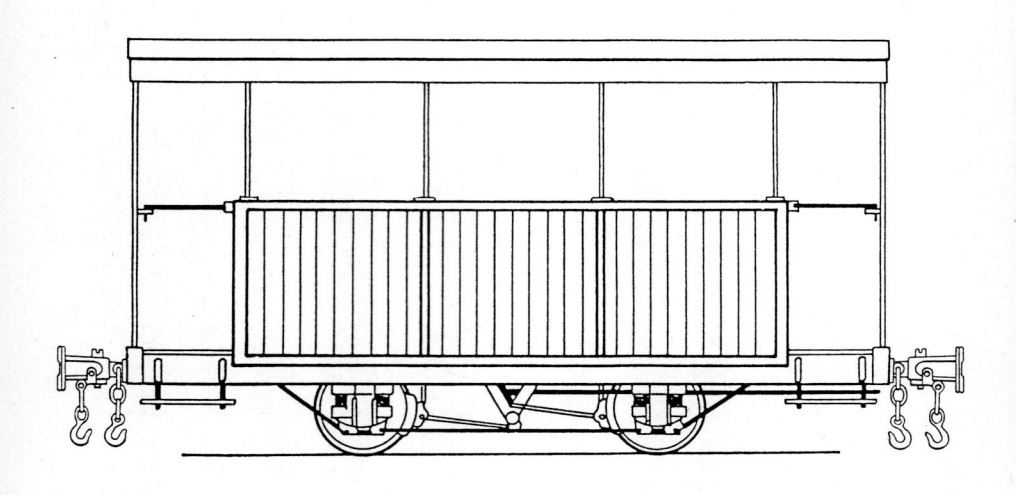

1891 Small end balcony open saloon, late form with curtains removed and ends boarded.
Scale 7mm = 1ft

Small open carriage with end balcony

The third non-standard coach was quite different from anything else on the line. It was particularly narrow with a width of only 4ft 6in giving a restricted carrying capacity of some ten people comfortably seated. This led to its rather limited use, the Company pressed it into service when no other stock was available. Although it has been suggested that the coach was completely open above the waist when it was delivered, a drawing from the *Midland Railway Carriage and Wagon Co.*, No. 2252, dated 6th March 1891 shows it fitted with a roof supported by iron stanchions and this was probably the situation when it was new. The ends of the carriage were open with steel railings but such an arrangement was clearly impractical in wet weather and full length panelling was soon fitted, this was to be carried throughout its life. Further protection to passengers was afforded by the fitting of canvas screens which could be drawn across the sides.

Passengers entered this coach not by the usual compartment door but by an open platform at each end. It is clear from drawing 2252 that access to each end platform was originally planned to be of width 2ft 6in. but that this was altered before construction to give a reduced width of 2ft. In overall appearance the end balcony carriage had a distinctly 'colonial' character, it was probably built for an overseas narrow gauge railway and only by a chance event, perhaps a failed order, did it become available at exactly the time when the Tramway was urgently seeking passenger carrying vehicles.

Dimensions of the open carriage with end balcony

Length	- overall	14 ft	6 in	Height - rail to roof	8ft	0in
	- over frames	10	6	- internal	6	0
	- internal	10	0	Floor level	2	0
Width	- over mouldings	4	6	Wheelbase	5	6
	- internal	4	0	Wheel diameter	1	9

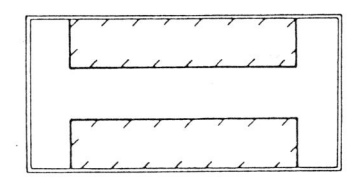

Seating for 10 passengers in small end balcony open saloon

1892 Series third class compartment carriages

Two closed coaches were delivered to the Tramway in 1892 by the Midland Railway Carriage and Wagon Co., each one cost £88. The fully panelled body with two doors on each side was mounted on a four-wheeled underframe. Individual footsteps were fitted below each door, on one side only, the north or "valley" side, grab handles were mounted adjacent to each door handle and ventilators were fitted over each door and in the roof. One end of the carriage was fitted with four steps and a handrail in order to allow access to the roof with its centrally mounted oil lamp, the other end had a parking hand-brake.

The conventional hinged door carried a droplight operated by a central leather strap suitably impressed with the GVT monogram. It allowed the passenger to enter a comfortable compartment with upholstered seating for eight people, this was divided from the second compartment by a three-quarter height partition panelled in brown varnished wood. The upholstery was of buttoned maroon cord cloth, and was stuffed with horse hair. While the interior was quite plain without mirrors or luggage racks and had a white finished roof carrying a central lamp globe, it was enlivened by curtains fitted during the early years of service. Each carriage was based on an undated and unsigned drawing prepared by the coach builders but during production several changes were made. The proposed left hand opening door became right hand opening, the straight grab handles were, in practice, curved and most important the wooden slatted seats were modified to much more comfortable upholstered seats.

The author David L. Davies was a boy in 1932 and recalls in his book 'The Glyn Valley Tramway' a journey in one of these coaches on a dark winter evening. He graphically describes the close heavy atmosphere with the mixed smells of lamp oil, manure on farmers clothing and tobacco smoke.

Dimensions of the 1892 series third class closed carriage

Length	- overall	14 ft	3 in	Height	- rail to roof	8ft	5in
	- over frames	11	7		- internal	6	5½
	- internal	11	1		- floor level	2	0
Width	- over mouldings	6	11	Wheelbase		5	6
	- frame	5	10	Wheel diameter		1	8
	- internal	5	4				

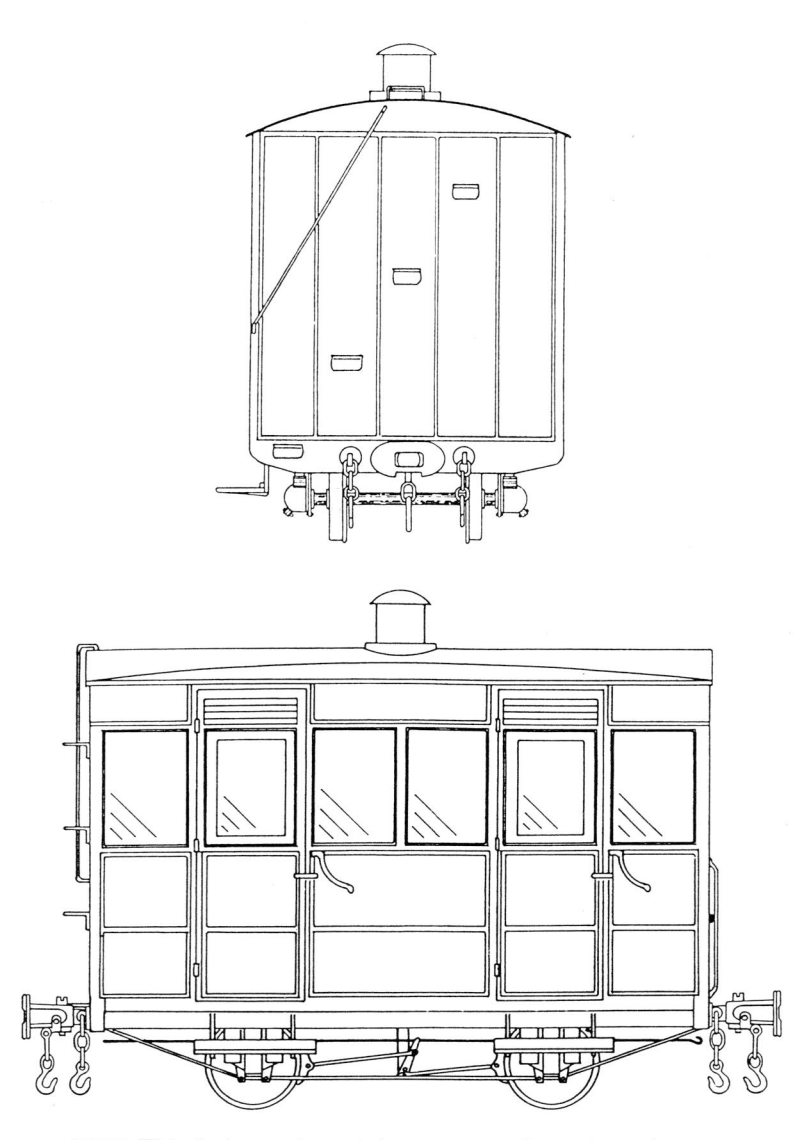

1892 Third class closed, two compartment carriage.
Scale 7mm = 1ft

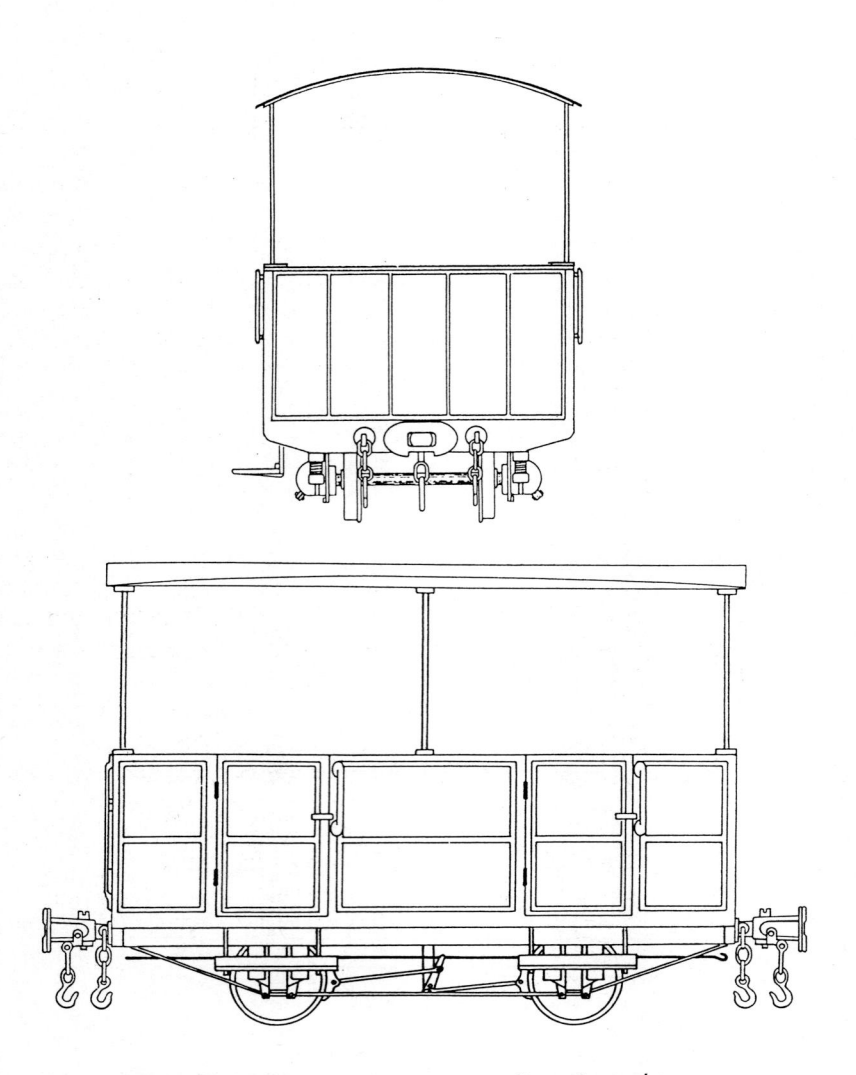

1893 Third class open, two compartment carriage, early form with open ends and corners.
Scale 7mm = 1ft

Open third class carriages

Early in 1893 six open coaches were delivered to Chirk and entered service on the tramway, they were built by the same company as the closed vehicles just described and were the result of an order placed during the previous year. The rake of coaches cost the princely sum of £360 in total, £60 each. In style and design, they were closely similar to the 1892 series third class closed carriages with two doors in each side and slatted wooden seats arranged in two compartments. The centre partition extended to waist level and was the same height as the seat backs. Sixteen, one and a half inch wide laths were used together with heavier two by three inch wide members at the front of the seat and the top of the seat-back. The panelled sides and ends were carried only to waist height leaving the roof to be supported by six iron stanchions. Four inch tongue and groove boarding was used to construct the roof and rain strips were fitted.

Access to the coach was facilitated by the long brass grab handles, these and the brass door handles were manufactured by *Fosters* of Birmingham. It is probable that the interior livery was brown varnish which was later modified to brown paint although one source states that latterly green paint was used to match the exterior.

While these coaches were intended primarily to be used for the summer tourist trade in the valley, the accommodation was rather spartan in cold or wet weather and would have been quite unsuitable during the winter months. The company recognised this and after a few years made appropriate modifications in the interests of greater passenger comfort. Around 1910, wooden end screens and four small side panels at each corner above the waistline were fitted. Both were formed from three inch tongue and groove boarding. After these modifications only a single iron stanchion was visible on each side, it was three feet and one inch high and one inch thick.

Dimensions of the 1893 series third class open carriage

Length	- overall	14 ft	3 in	Width	- over mouldings	6ft	11in
	- over frames	11	7 *		- frame	5	10
	- internal	11	1		- internal	5	4
Height	- rail to roof	9	6	Wheelbase		5	6
	- internal	6	5½	Wheel diameter		1	8
	- floor level	2	0				
	- of side panel	2	9				

* In 1980 members of the Glyn Valley Tramway Group took measurements of a surviving open coach at Llwynmawr confirming the figures given in the table however a length of 11ft 9in over frames was obtained.

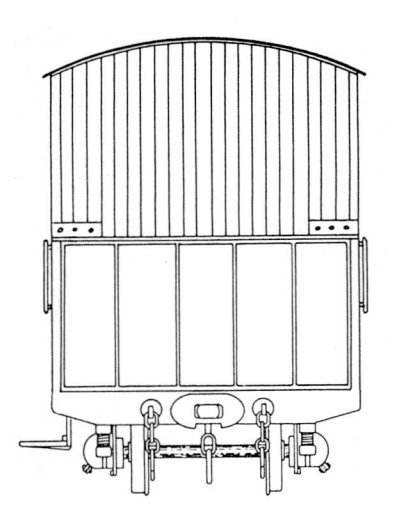

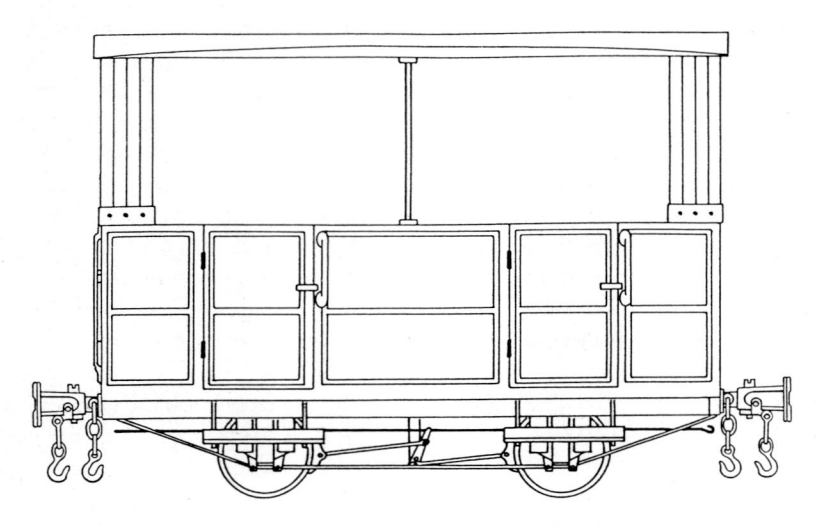

1893 Third class open, two compartment carriage, late form with boarded ends and corners.
Scale 7mm = 1ft

Although invisible, the four corner stanchions were retained behind the panelling. Decorative beading on the coaches was one and half inches wide with rounded ends. Footboards which measured two feet three inches long and nine inches wide were fitted only on the north side of the coach and a lever operated parking brake was mounted at one end. In modified form, the open coaches enjoyed great popularity during the summer months and a number of photographs taken during the twenties and thirties show them filled with eager tourists.

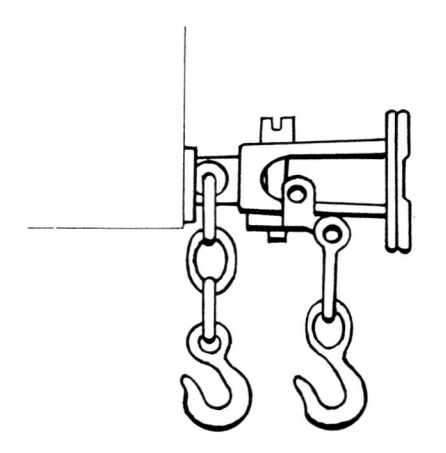

Coach buffer showing centre coupling hook on right and one of the two side safety chains and hooks on left. The pin which passed through the coupling shank is down in place on the top centre of the buffer

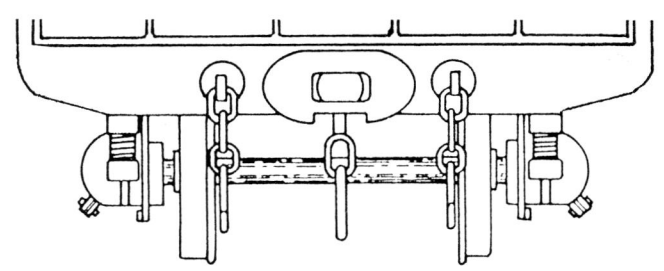

Coach buffer beam showing centre buffer and centre coupling hook with safety chains and hooks on each side just inside the wheels

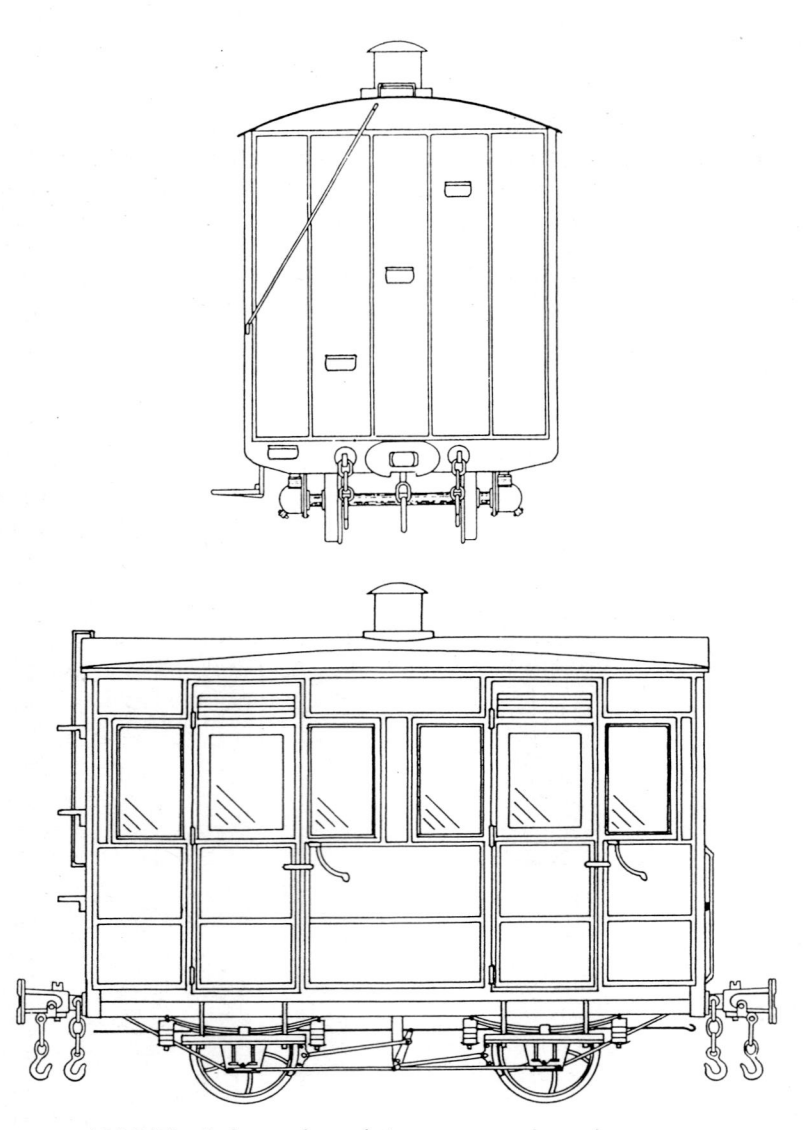

ca. 1893 First class closed, two compartment carriage.
Scale 7mm = 1ft

The closed first class carriage

It was without doubt the growing confidence of the Directors in the early 1890's that led them to order for the company a carriage designed to provide luxurious accommodation for the discerning traveller. While it is difficult to understand how the purchase of such a vehicle could be justified on strictly economic grounds, it would provide a highly visible symbol of the strength and prosperity of the company. Only a single carriage was commissioned, probably in 1892 although the absence of a date on the design drawing No. 2347 of *The Midland Railway Carriage and Wagon Co.* and the failure of order books to survive, makes this date uncertain.

It was delivered in 1893 and was a splendid vehicle in all respects, outwardly it was similar to the third class coaches which were to be built in 1901. The fully panelled sides were fitted with two doors each hinged at the left. The roof carried a single central, ventilated lamp pot access to which was provided by four steel footsteps and a handrail on one end. The other end was fitted with the handle for a parking brake. Passengers used the footsteps below the doors and the curved brass handrails on entering the coach. There were no footsteps on the south side of the vehicle. The door with its padded interior and top ventilator carried a droplight operated by a leather strap.

The interior can only be described as opulent, it was quite different from anything else on the Tramway. The seats had sprung cushions and backs under horse hair padding and were button-upholstered with corded blue cloth. Matching padded elbow rests were fitted below the side windows. The doors, seats and elbow rests were generously trimmed with twisted cord. The upholstery was offset by rich mahogany panelling while the roof panels with adjacent sides and ends were lined with white lincrusta wax cloth. The roof was decorated with gilt mouldings. The interior was generously fitted with twin mirrors placed over the seatbacks at each end and below the luggage racks. Draw-down blinds were provided over the windows and door droplights. The two compartments, each with seating for six passengers, were divided by a three-quarter height partition and this allowed the single colza oil lamp to illuminate the whole coach.

Dimensions of the closed first class carriage

Length	- overall	14 ft	3 in	Height - rail to roof	9ft	6in
	- over frames	11	7	- internal	6	5½
	- internal	11	1	- floor level	2	0
Width	- over mouldings	6	11	Wheelbase	5	6
	- over frame	5	10	Wheel diameter	1	8
	- internal	5	4			

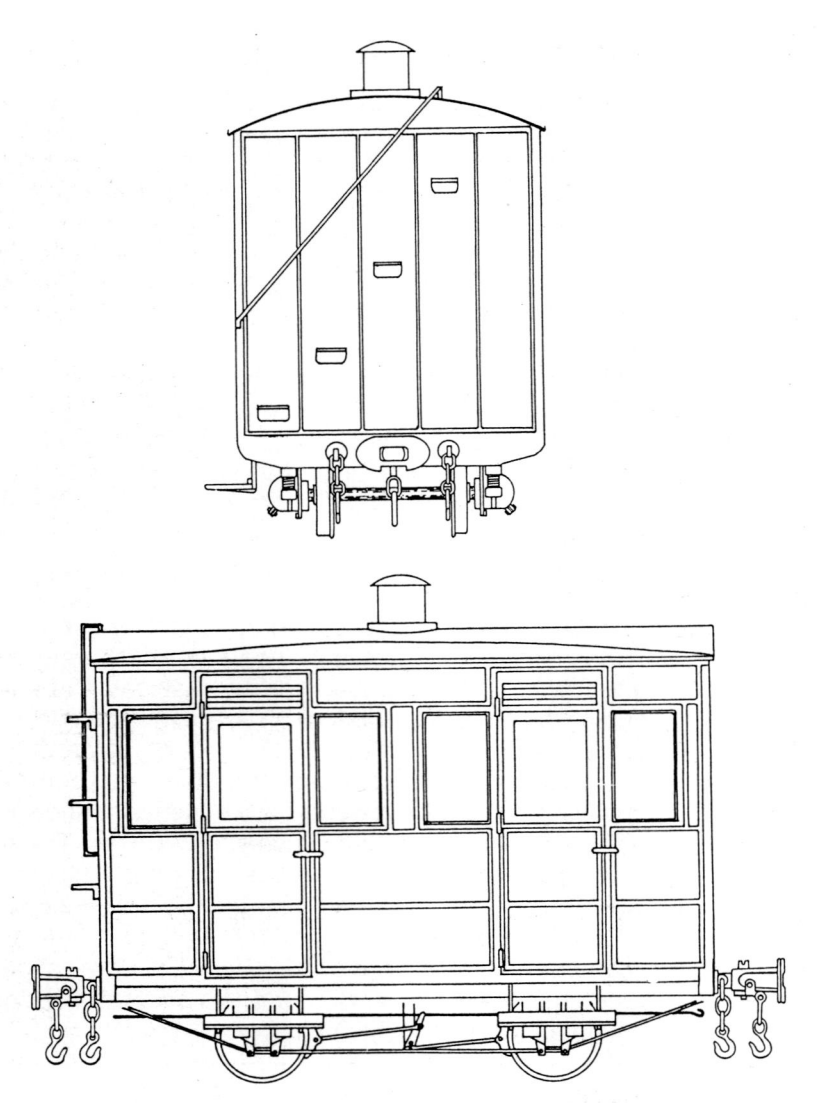

1901 Third class closed, two compartment carriage.
Scale 7mm = 1ft

1901 Series third class compartment carriages

Two closed third class coaches were delivered to the Company by the Midland in 1901, they were essentially the same in outward appearance as the first class coach although slightly higher in the body. They appear to have been purchased as replacements for the clerestory coach and the centre-door closed saloon and soon displaced these two from regular service, by 1904 they had been officially downgraded to the status of "workmens coaches".

The new coaches differed from the 1892 series closed carriages in the design of the side panels and windows. The 1901 series had panels inserted beside the windows at the two ends of each side, a similar narrow panel was placed between the windows in the centre of the side. Thus the windows were narrower in width although of the same height as in the 1892 series.

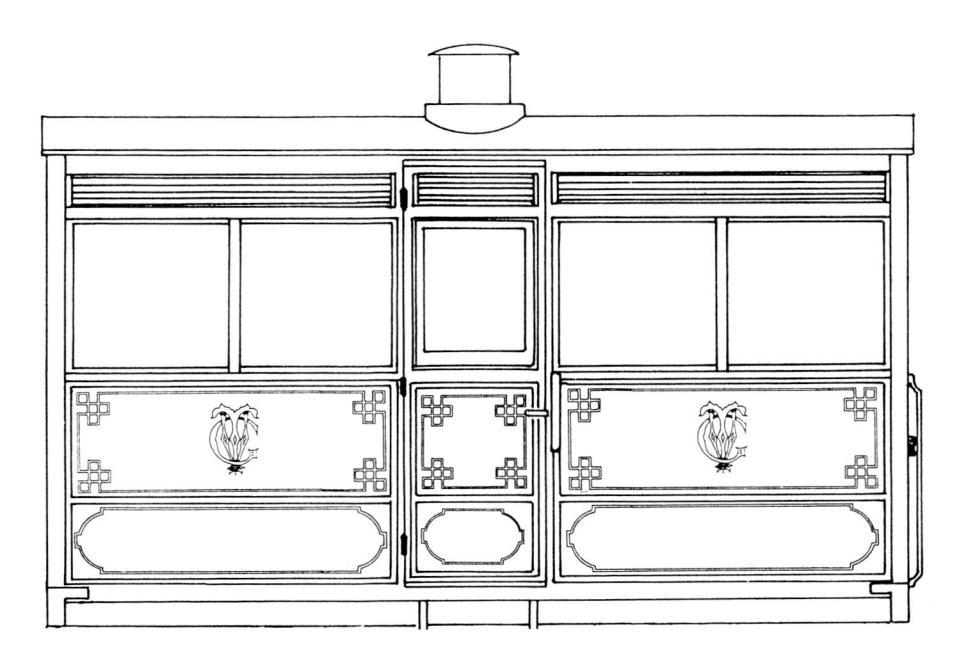

Centre door closed saloon in first pattern, fully-lined livery carrying monogram with predominant 'T'. The clerestory closed saloon carried the same livery pattern (1891-ca. 1913)

A Chirk bound excursion train is about to leave Glynceiriog Goods Yard, every type of coach used on the Tramway is shown, 28th August 1926. (The late HC Casserley).

Three open third class carriages together with the small open carriage stand off the rails in Chirk Yard, 1936. (MJ Higgins collection).

A Glynceiriog bound passenger train stands at Pontfadog, the coaches carry the first pattern lined livery with GVT monogram, ca. 1905 (Lens of Sutton).

A rake of coaches in the loop near Chirk Station, the pale colour of the sides of several vehicles is clearly visible.
(The late HC Casserley).

Curved brass grab handles replaced the earlier pattern straight handles and on the end of the coach the bottom footstep was fitted rather higher inside the panelling. The seating arrangements and the layout of the interior remained the same as in the 1892 series.

Dimensions of the 1901 series third class closed carriage

All dimensions were the same as those given earlier for the 1892 series closed third class carriage.

The Tramway ran close to hedgerows for much of its length, the danger that this posed for passengers who leaned from the windows on the south side of coaches was obviated by fitting two vertical iron window bars to the door droplights on this side. It is probable that all coaches with droplights had the bars in place by 1895 although they are not shown in any original coach drawings.

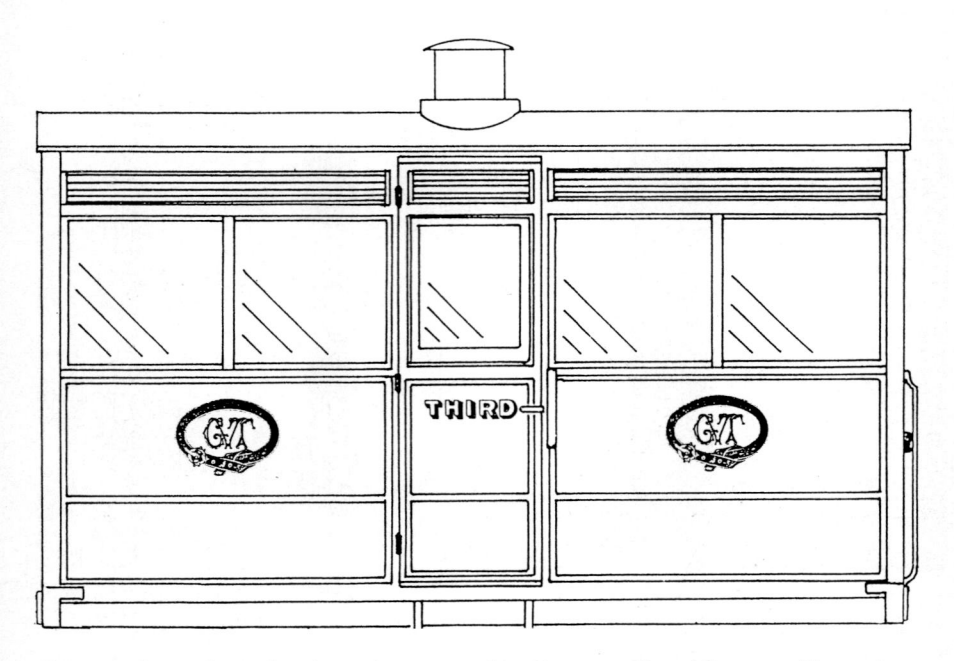

Centre door closed saloon in second pattern, unlined livery with garter monogram and class designation (ca. 1913-ca. 1925)

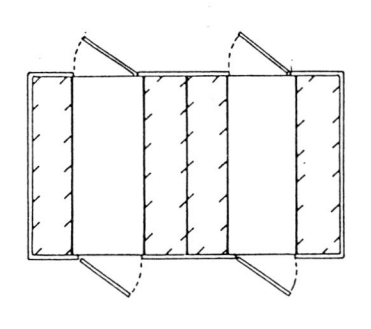

Seating for passengers in 1892 and 1901 closed coaches, 1892 open third coaches. The first class carriage used the same arrangement but seated only 12 passengers as arm rests were fitted between individual seats

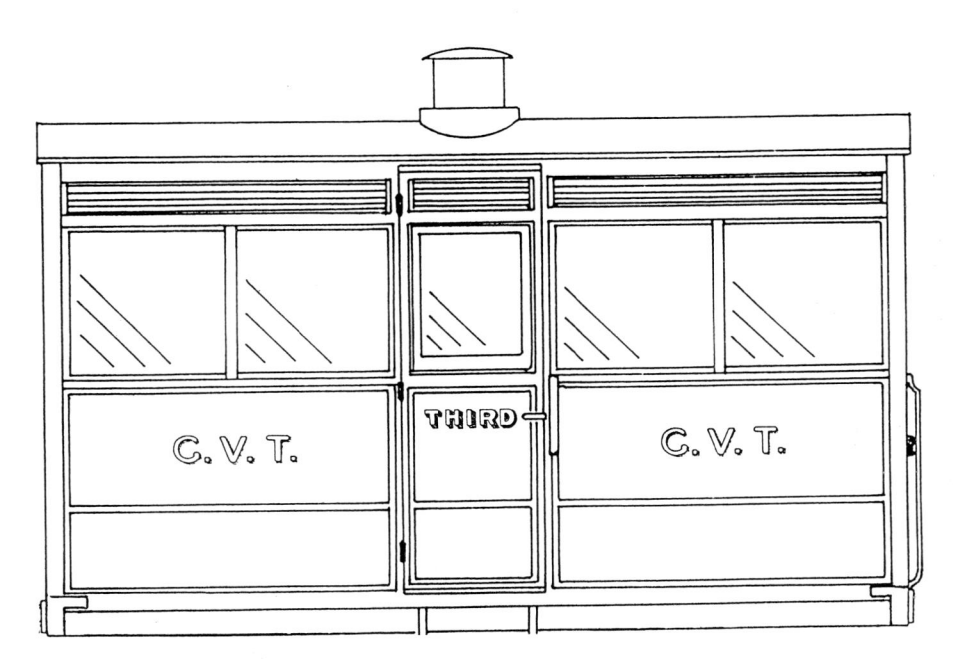

Centre door closed saloon in third pattern, unlined livery with thin lettering (ca. 1925-1935)

Coach Livery

The body of each coach on the Glyn Valley Tramway was painted in medium green, this was a colour similar to Great Western Railway green, it was sometimes described as holly green. Frames and other features were picked out in ivory which has also been referred to as cream or buff. Early photographs ca. 1896, suggest that it was very light, possibly even white, over-coated with clear varnish. If this was the case then the colour would have weathered to cream or buff. The roof was originally painted white and this darkened in normal service to various shades of grey. The lead-based white paint which was in common use at the time was particularly susceptible

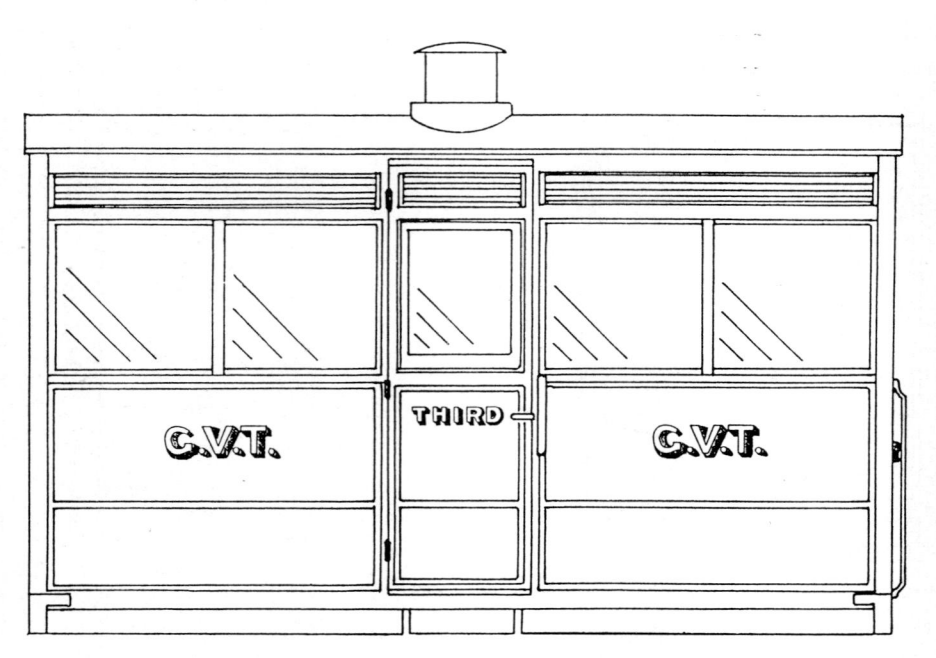

Centre door closed saloon in fourth pattern, unlined livery using thick, heavily shaded, lettering (ca. 1925-1935)

to darkening in the presence of sulphur fumes such as would be produced by a coal burning locomotive.

Livery styles changed a good deal during the long life time of the coaches tending to become much more simple. The earliest livery to be used was an elegant and elaborate lined decoration on the side panels using quarter-inch or half-inch wide gilt lining. On the closed and open third class coaches and the first class coach, the horizontal row of panels including the door panels immediately below the windows, or waistline on open coaches, were decorated with a quartered style of lining as shown in the drawing. The second, bottom, row of panels was given a simple lined finish with indented corners.

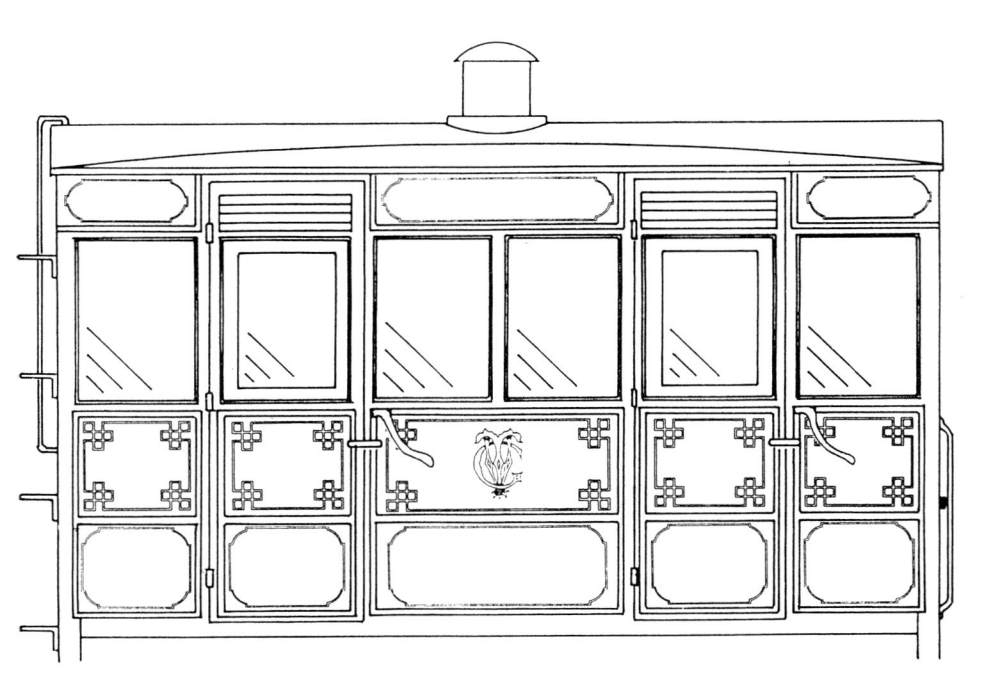

1892 Closed third class carriage in first pattern, lined livery with monogram showing predominant 'T' (1891-ca.1913)

31

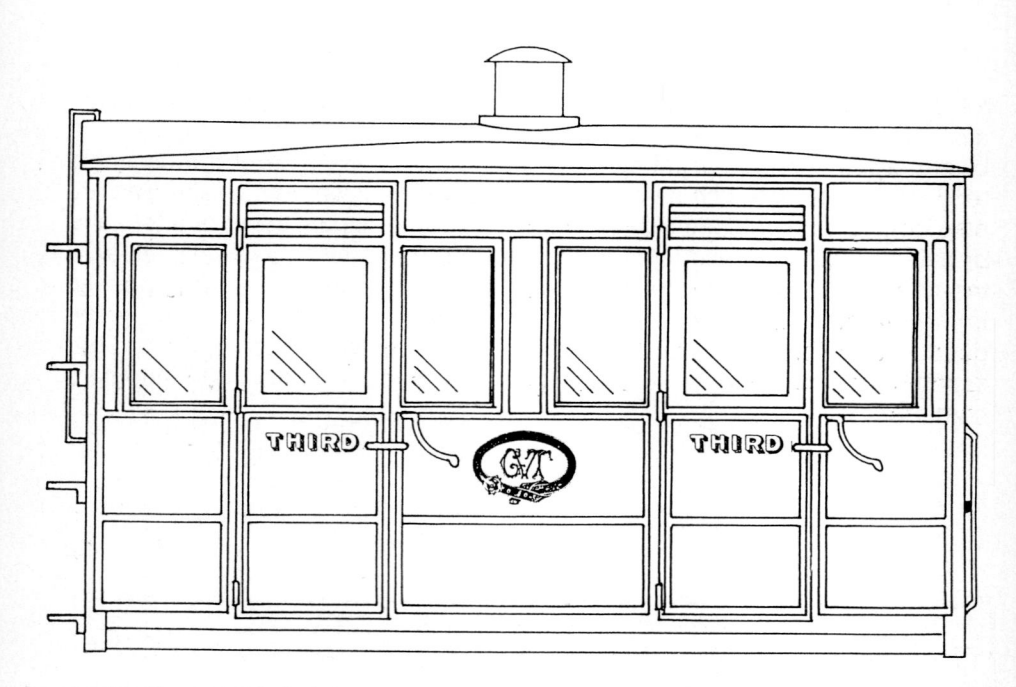

1901 Closed third class carriage in second pattern, unlined livery with garter monogram and class designation. The 1892 closed third class carriage had the same livery (ca.1913- ca.1925)

The bold gilt GVT monogram was emblazoned on the top centre side panel of the closed and open four door coaches. The centre door coaches carried the monogram twice on each side in the centre of the large side panels below the windows. There were two distinct styles of monogram, one with the "T" predominant which was used on the first and third closed coaches, the second style had a predominant "V" and was applied to the 1893 series open third class coaches.

The end balcony open carriage carried no livery at this time, it was unlined and had no monogram. Class designations were not carried.
The running number was painted in black on the right hand side of the solebar. Early lining styles and monograms are clearly shown in photographs 99 and 103 from the GVT Group reference collection.

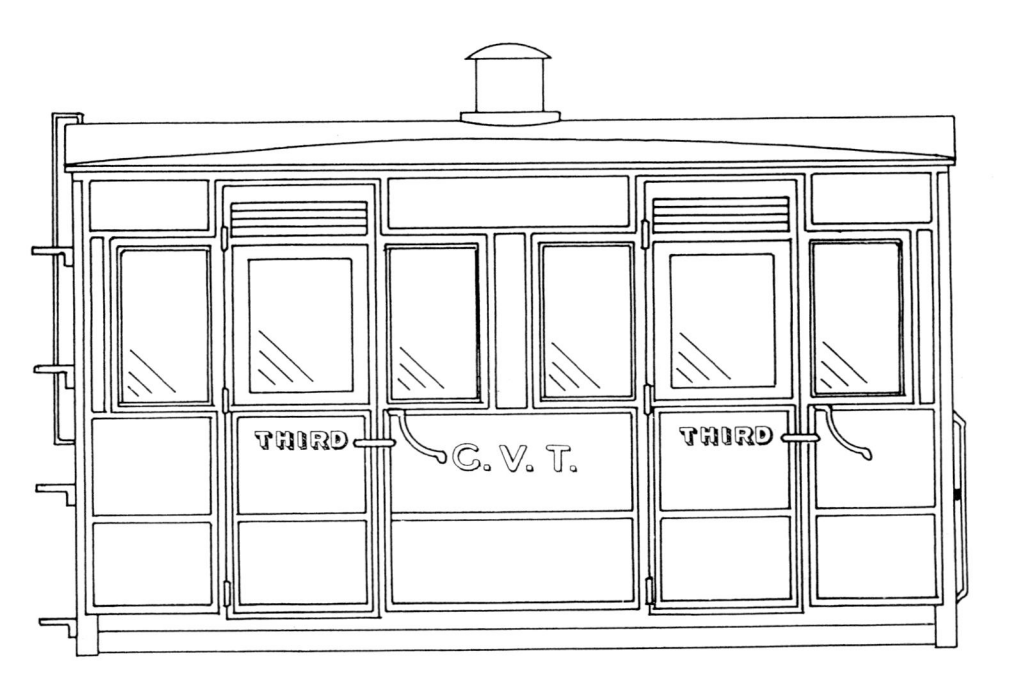

1901 Closed third class carriage in third pattern, unlined livery with thin lettering (ca. 1925-1935)

Below: 1901 Closed third class carriage in fourth pattern, unlined livery with thick, heavily shaded lettering (ca. 1925-1935)

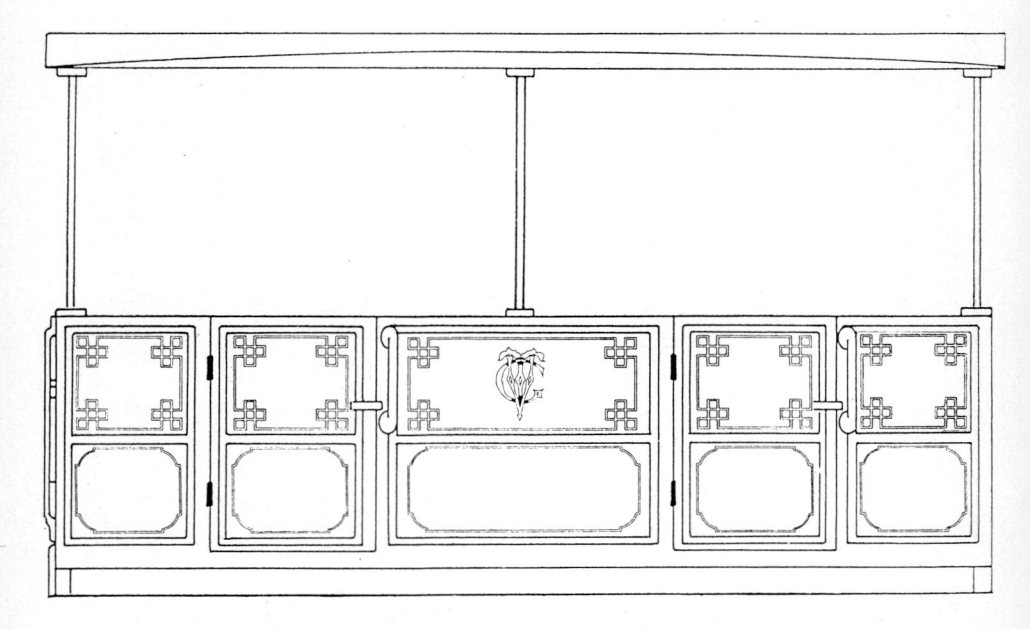

1893 Open third class carriage with open ends in first pattern, lined livery, the monogram has a dominant 'V' (1891-ca.1910)

A new unlined livery was used from about 1913, it incorporated a third style of monogram set within a vermilion garter, it replaced the earlier forms. The coaches continued to be painted in medium green and ivory. The monogram was applied to the top centre side panels of closed and open four door carriages while it appeared twice on each side of the centre door vehicles. Class designations of "FIRST" and "THIRD" were formed in two-inch high gilt letters shaded red, these were applied as appropriate to the doors of open and closed coaches and to the left and right side panels of the small open carriage. The transfers were supplied by *Therne and Sons Ltd.* Carriage numbers were of two-inch high gilt numerals shaded black and were affixed to the right hand corner of the solebar.

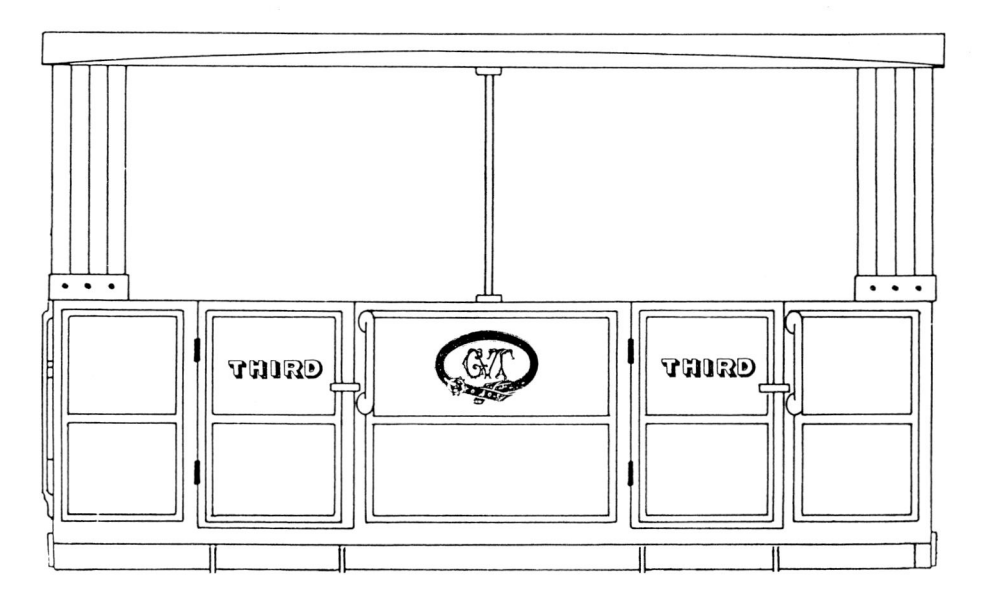

1893 Open third class carriage with boarded ends and corners carrying garter monogram and class designation (ca.1913-ca.1925)

A modified style of unlined livery was introduced around 1925 in which the monogram and garter were replaced by the letters "G.V.T." in six-inch high, hand painted, gilt or yellow letters shaded red. A variation was used on the centre door saloon and clerestory coach where the letters were nine inches high with the running number now painted in yellow on the right hand corner of the solebar. Initially the lettering was of high quality with clear, thin, well balanced lines. When repainting became necessary a much lower quality was adopted, the letters became much thicker and less well formed, the shading was very heavy. This style endured until closure.

All of the closed carriages were originally designated "SMOKING" and "NON SMOKING" and the appropriate indication was carried in gilt lettering on one window of each side.

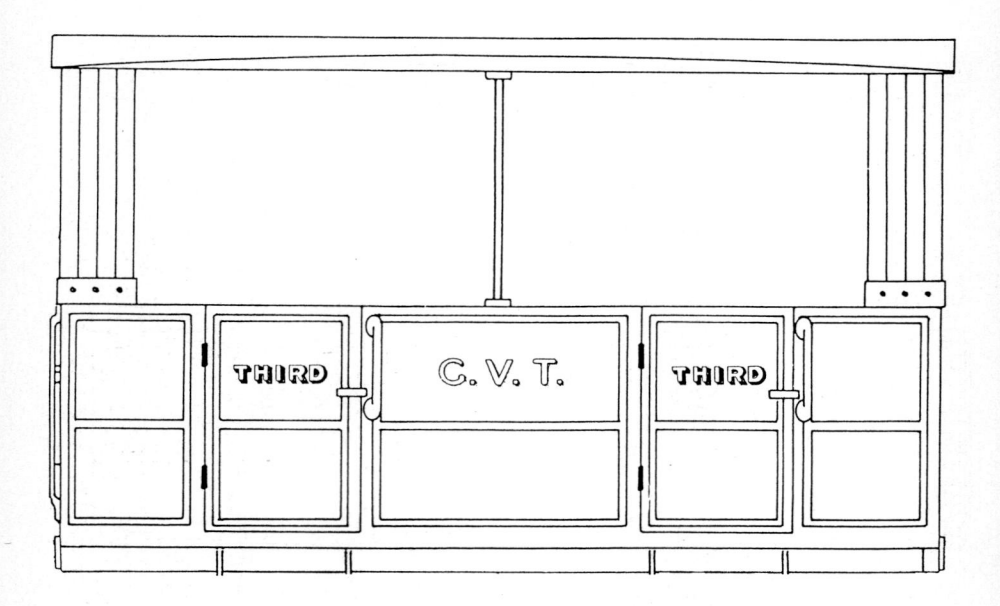

1893 Open third class carriage in third pattern, unlined livery with thin lettering (ca.1925-1935)

Some specific details of individual coaches will now be discussed. Very little is known about the livery of the horse tramway stock. In photographs, the colour appears to be lighter than the green used latterly. The paint may have been dark green when originally applied but fading in service to a lighter blue-green. In the steam era the coaches appeared in different colour patterns under lined and unlined livery. The clerestory coach in the earliest lined style had the ends painted green with ivory corner posts. The sides, panels and droplights were green with the window and door frames in ivory. When in the later unlined livery it had the ends and sides painted green, the window and door frames were finished in ivory.

The centre door closed saloon when in lined livery had the main end panels painted green with ivory tops and corner posts, later this was modified to

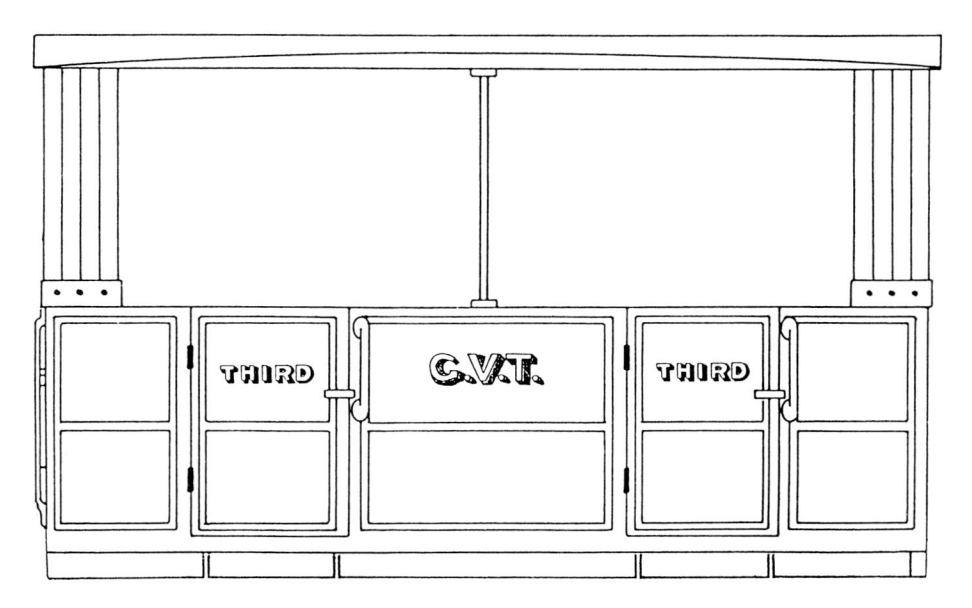

1893 Open third class carriage in fourth pattern, unlined livery with thick heavily shaded lettering (ca.1925-1935)

plain green ends. The side panels, ventilator louvres and droplights were green with the rest of the side painted in ivory. The later unlined style left the ends painted plain green with the side panels, ventilators and droplights also in green. The remainder of each side was ivory above the waistline and green below it.

The first class coach and the 1901 series closed third class coaches in lined form were painted green on the end panels with ivory tops and corner posts, later the whole of the end was finished in green. The side panels including the window level panels and the droplights were green with the remainder of the side in ivory. In unlined form, the coach ends, side panels, ventilators, droplights and corner posts were green. The window level panels were ivory with green beading, the rest of each side was ivory above the waistline and green below.

The 1892 series closed third class coaches first appeared in lined livery with the main end panel painted green with an ivory top and corner posts, later the end was changed to plain green. The side panels and droplights were green with the remainder of the side finished in ivory. Unlined livery consisted of green coach ends with side panels, ventilators and droplights in the same green, the rest of each side was painted ivory above the waistline and green below it.

In their earliest form with open ends, the 1893 series open third class coaches were out shopped with the main end panels painted green with ivory frames.

The side panels were green with the remainder of the side painted ivory. Early livery details of the small open end balcony saloon are sketchy, a photograph shows the coach in a train at Glyn Ceiriog station ca. 1900. The side panel is dark green with its edge beaded around all four sides in ivory. Surprisingly, the normal lining and monogram was not applied to the vehicle during its early life. Latterly the sides and ends were painted in unrelieved green.

The ironwork on all coaches was painted black and from about 1913 the roofs were treated with black bitumen, this gradually weathered to dark grey in service. Several photographs taken over a period of years from about 1929 until the closure of the line in 1935, show the clerestory coach and the centre door closed saloon in what appears to be a lighter shade of green paintwork. It is our opinion that these vehicles were not repainted during these last years of operation and the difference in colour was due to weathering and fading rather than to refinishing in a lighter shade of green. Paints used at that time were usually lead-based, this would lead to gradual fading of the original colour to a blue-green shade and eventually to dark grey when left open to the elements for a period of years.

Brakes, running gear and couplings

The Parliamentary Bill which authorised the construction of the tramway using steam powered locomotives was passed in 1885 under the title of the Glyn Valley Tramway Act. The Bill stipulated that: *"Every carriage used has to be constructed as to provide for the safety of passengers and be fitted with a brake which shall be connected with the engine and applied by the driver when necessary."*

This requirement was met for the opening of the line to passenger traffic by connecting the steam brake on the locomotive to manual brakes on the carriages by means of chains under the carriage floors. When the locomotive brake was applied, the brake hangers tightened the chain connections and thence mechanically applied the carriage brakes.

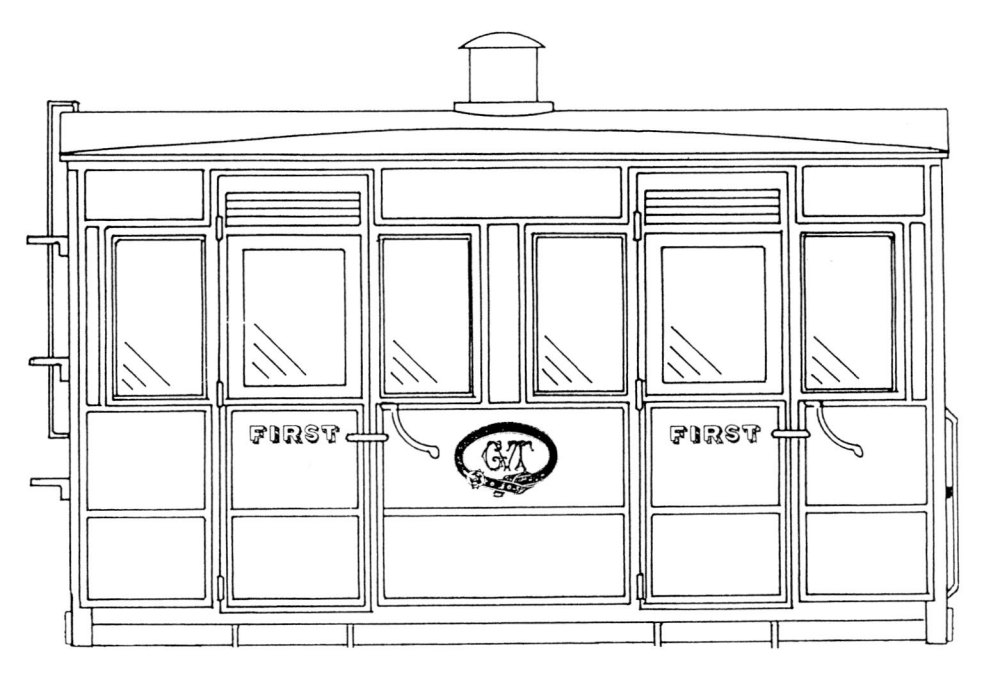

First class carriage in second pattern livery with garter crest and class designation. Other livery patterns for this coach matched those of the 1901 closed third class vehicles (ca.1913-ca.1925)

The first steam passenger train has just arrived at Glynceiriog from Chirk. The clerestory coach is in its early condition with end mounted lamps, full length door and a shallow clerestory roof. The second coach is an ex-horse-tramway vehicle, 16th March 1891.
(MJ Higgins collection).

A restored GVT carriage in service on the Talyllyn Railway, August 1989. (MJ Higgins collection).

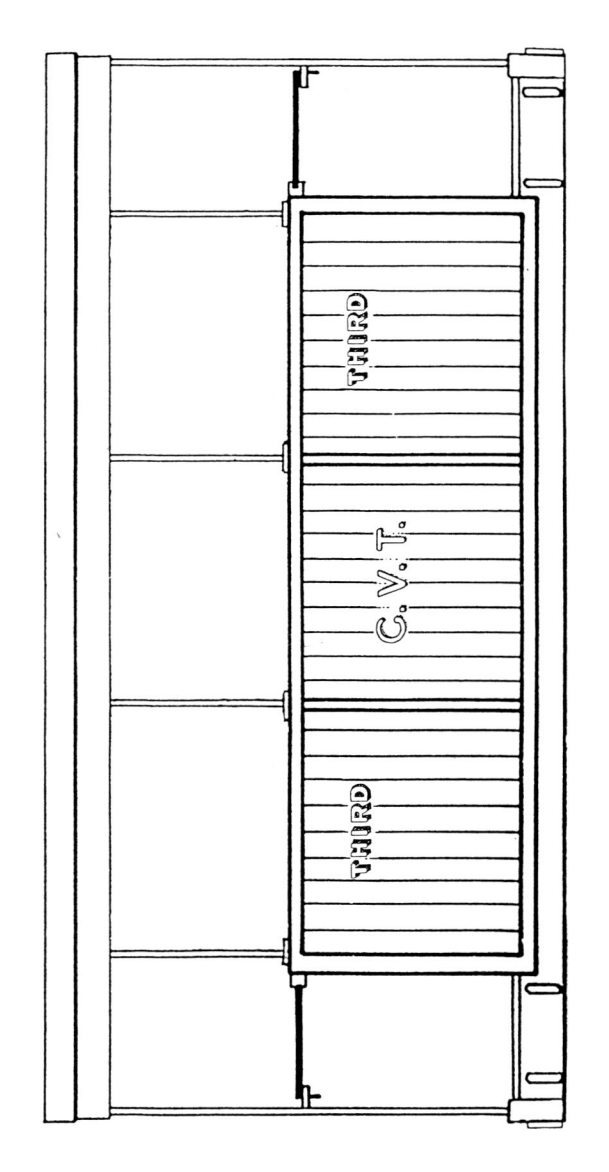

1891 Small end balcony open saloon in third pattern livery. The vehicle was plain and unlined until it received this decoration (ca.1925-1935)

This concept was of only limited success under normal operating conditions as the chain links broke under tension on sharp curves, it fell into disuse by 1914 although unfounded rumours suggest that it was abandoned in the 1890's. A single lever parking brake was fitted to each of the thirteen coaches with the exception of the small open end balcony saloon. The brake lever was fitted with a peg and the curved backplate had a series of matching holes drilled in it. The brake was applied by pushing the lever downwards as far as possible then the adjacent hole was engaged.

Each coach was turned so that the brake lever was positioned at the north side of the vehicle projecting towards the main road. After several years in service, the small open saloon was fitted on one end balcony with a pillar mounted handwheel which operated a parking brake by means of a screw mechanism accessible from inside the coach or from outside through a sliding panel in the end of the vehicle.

Coach underframes were of timber construction in every case, they carried sprung axleboxes. The first class coach was unique in being suspended with leaf springs on each axlebox. Laminated springs were fitted to each axlebox on the closed third class coaches. The axlebox patterns varied slightly between first and third class coaches while the 1892 series third class vehicles were built with S pattern axleboxes. The six open third class carriages were mounted on underframes of a different pattern with coil springing, tie bars and a modified axlebox. This pattern was followed by the clerestory coach, the closed centre door saloon and the small open end balcony coach.

Sprung centre buffers were fitted on the coaching stock and the individual carriages were coupled together by means of a coupling link, this was a short shank of forged iron with an integral loop at each end. The end of the coupling link was passed into an aperture in the buffing head and the link was secured by dropping a peg through the vertical hole behind the buffing head. A spare link was always carried and photographs show that it was hung under the corner of the coach underframe.

A side chain and hook was fitted on each side of the centre buffer, these were linked between adjacent coaches for safety in the event of a coupling link failure. In addition, a single link and hook was located under the buffer to permit the connection of the coach to the locomotive, goods van or wagons of various types since the coupling link could be used only for coupling between coaches or a coach and a brake van.

Wheels used were of one foot eight inches or one foot nine inches in diameter with six curved spokes of one and half inches diameter. The wheels and tyres were steel and the wheel width was 4 5/16 in., the wheels were carried on an axle of thickness three and half inches which tapered to two and half inches at the axlebox. The length of taper was five and half inches with a back to back measurement of 2 ft. 2 1/8 in. A comment should be made on the wheel diameter, it seems likely that measurements of one foot eight inches which were made during the years immediately prior to closure were a result of turning and reprofiling the tyres over a period of years to give a smaller size from the original one foot nine inches.

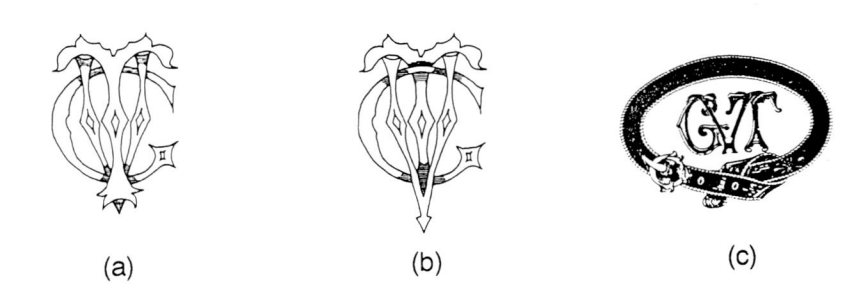

(a)   (b)   (c)

Monograms used on Glyn Valley Tramway coaches. (a) First pattern, 1891-ca.1910, used with fully-lined livery on closed coaches, "GVT" with "T" emphasised. (b) As (a) but used on open coaches, except end balcony open coach, "GVT" with "V" emphasised and elongated. (c) Second pattern, ca.1913-1925 with GVT enclosed in a garter, used on closed and open coaches but not on the end balcony open.

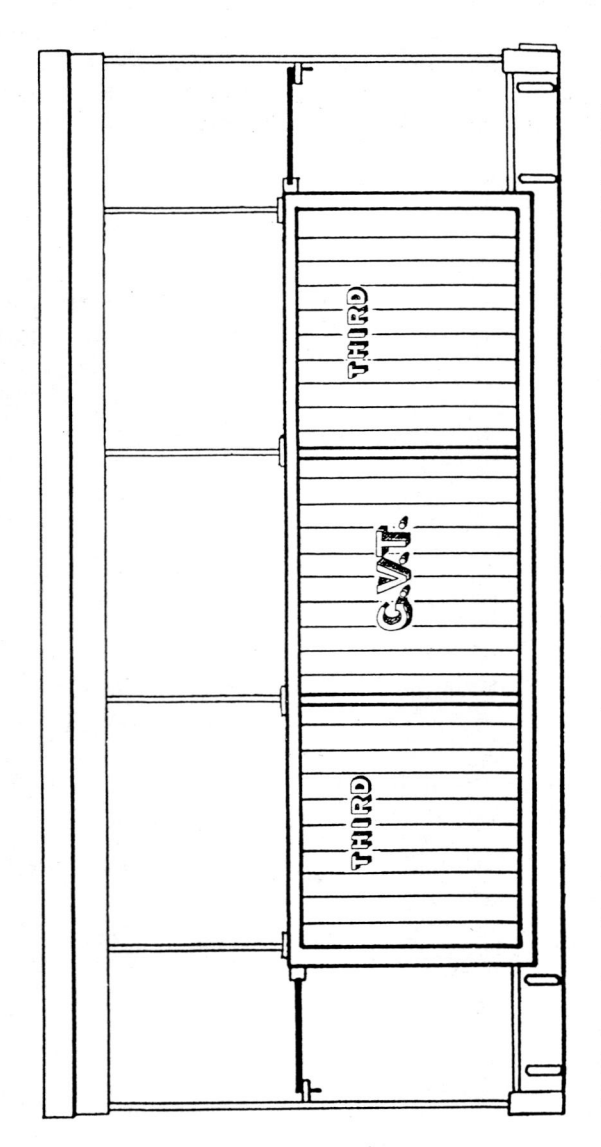

1891 Small end balcony open saloon in fourth pattern, unlined livery (ca.1925-1935)

Closure and dispersal of the stock

Passenger services on the Glyn Valley Tramway were withdrawn on Thursday 6th April, 1933 and the line was closed to all traffic on Saturday 6th July 1935. Photographs taken after the closure show the coaches stored in the open sidings at Chirk, the poor condition of the paintwork and broken windows suggest that they were unmaintained. In 1936, the wheels and running gear were removed and the coach bodies sold at auction. The fate of these dismounted bodies is far from clear, most seem to have decayed and collapsed in the unrecorded obscurity of farms or gardens in the Ceiriog Valley.
However rumour and comment have enabled some to be traced and two have survived to the stage of preservation. According to an historian, whose family was acquainted with the contractor responsible for the demolition of the line, the clerestory coach, the closed centre door coach and the open small saloon were shunted after closure to the north end of the Chirk yard beyond the tar tips. Local wisdom informs us that any stock moved to that area was burnt to destroy the timber and allow the metal parts to be salvaged as scrap which was subsequently removed by rail. The clerestory coach was at this time in appaling condition with a damaged and sagging clerestory roof and all windows missing, it would have been an obvious candidate for scrap.
The first class coach was traced initially to Mr Moss Edwards, a well known resident of Llwyn-y-cil near to Chirk. He sold it to a garage proprietor at Chirk in the 1950's and according to an article in the *"Oswestry Observer"*, *"while there, the owners of the Talyllyn Railways, one of the few narrow gauge railways now existing, had made enquiries with a view to purchase"*. The change of ownership did not however take place and in 1953 the Rev. W.C. Dicken, vicar of Chirk bought the coach for use as a hen house. It was spared this demeaning assignment by the vicars children who persuaded their father to let them have it as a play house. It served this purpose for some three years until early 1956 when Mr Scholes of the British Transport Commission Historical Department wrote to Chirk Parish Council about the possibility of restoring and preserving it. The letter appears to have evoked a negative response since no action was taken. The *"Border Counties Advertiser"* picked up the story and asked that a permanent home be found for it in Wales. This wish was fulfiled later in the same year when a member of the Talyllyn Railway Preservation Society purchased the coach and generously presented it to the Society.
In 1975 members of the Glyn Valley Tramway Group traced an open third class coach to a farm at Llwynmawr near Dolywern and photographed it. A second open coach was stored nearby but had been lost by the early 1970's.

The photographs, GVTG No. 103 and 104, show the coach as essentially complete with some traces of faded, weathered green paintwork still remaining after forty years of neglect and the legend "G.V.T." was still proudly displayed on the side panels. However by 1980 the roof had collapsed and the coach disintegrated in the following years. The ironwork from it, including the mechanical brakes, have been salvaged and stored, two side panels have been restored as have the door handles and grab handles. The latter are displayed along with a number of other artifacts and a photographic collection in the Glyn Valley Hotel, Glyn Ceiriog. The nearby Ceiriog Memorial Institute has a small range of relics including a set of curved grab handles and a droplight strap. A further set of grab handles from an open third class coach have been preserved privately.

Some fragmentary information is available on other vehicles, it has been said that two open third carriages were purchased at the 1936 auction for five shillings each (twenty five pence) by a Dogelley company for use by their steam traction engine crews. A 1901 series closed third coach was located at a garage in Chirk in the early 1960's but it has since disappeared. A lady in Glyn Ceiriog had an open third coach stored in her back garden for many years but in 1975 she broke up the body and underframe and burned the remains. A closed third class coach found its way to the blacksmith in Llwynmawr where it was located for some years around 1962 before being demolished but the window frames and windows have been saved. An 1892 series third class carriage was in use as a hen house on a farm near Glyn Ceiriog and was photographed about 1956, it has now disappeared.

One of the 1901 series third class vehicles did however meet a kinder fate, after being sold at the 1936 auction it was used as a shed or store on various farms in the Ceiriog Valley. After twenty years it was located on a farm at Ty-Isaf near Glyn Ceiriog and arrangements were made by the Talyllyn Railway to purchase it early in 1958. Surprisingly, it was found that the roof had been weather proofed by the application of a layer of concrete!

The dismounted body of a first class carriage at a garage in Chirk, 21st April 1950. (JG Vincent).

End view of the coach in the last photograph, 21st April 1950. (JG Vincent).

Restoration and preservation on the Talyllyn Railway

A very dilapidated first class carriage arrived at Towyn early in 1957 and after consideration, the Railway took the enlightened decision to restore it as far as possible to the condition and appearance it enjoyed while in service on the Tramway. However a number of modifications were essential in order to fulfil the operating requirements of the Talyllyn, these included the removal of the lamp housing which would have fouled the loading gauge and removal of the passenger footsteps. Ex-Ffestiniog Railway running gear has been used and twin buffers replace the centre buffer while the axle boxes use ball bearings.

Inside, the coach was divided by a waist high partition in place of the three-quarter height original, the luggage racks were removed and the seats fitted with foam rubber cushions upholstered with blue leather cloth. Interior panels have been finished with beige paper in place of the white lincrusta.

The coach has received a medium green and ivory lined livery with a white roof incorporating the 1913 style GVT monogram within a garter, it is lettered "FIRST" on the doors. The roof access steps and handrail on the coach end have been retained together with the parking brake and lever.

Restoration of the carriage was carried out in the Talyllyn Railway workshops at Pendre, Towyn by their own staff. The extensive programme of reconstruction needed was sympathetically achieved by expert craftsmen to give a wholly pleasing result. Some credit must nevertheless be reserved for the original builders, since the hardwood framework was found to be in good condition after more than forty years in service and a further twenty years of total neglect. It still retains the original underframe together with its old number, 14.

The third class coach acquired by the Talyllyn Railway was taken to *Hunts of Oldbury,* the railway and traction engine builders and restorers, for substantial rebuilding and modification before joining its sister in service on the railway. While the work was in progress, it was discovered that the double floor had a layer of insulating felt between the two skins. Parts of the original mechanical brake were found in the same place, this had been completely lost from the first class coach prior to its restoration. The same first class designation and livery was applied to this carriage as to no. 14.

It has received the Talyllyn no. 15. During its former existence on the Tramway it carried the numbers 6 or 7 for at least a proportion of the earlier years. Since the original restoration in 1958-59, the two coaches have been fitted with luggage racks and with lamp globes for electric lighting. Both vehicles receive regular routine maintenance and remain in excellent condition as a lasting memorial to the Glyn Valley Tramway.

The Glyn Valley Tramway Group

Readers who wish to learn more about the Tramway and the community it served may wish to contact the Group. Formed in 1974, it exists to collect and conserve evidence of the Tramway in the form of photographs, artifacts, ephemera and buildings. The Group researches the history and operation of the line, it provides an information service to enthusiasts and historians.

A small museum of GVT material is maintained in the Glyn Valley Hotel at Glyn Ceiriog, this is open to the public during licensing hours. The former Tramway station waiting room at Pontfadog has been purchased by the Group and restored to its original condition.

Membership Secretary: Bernard Rockett, Orchard House,
Manor Farm Lane, Oldbury, Bridgnorth, Shropshire, WV16 5HG

Coach numbering

The Glyn Valley Tramway Company assigned numbers to its coaching stock but did not retain the same numbers for all the vehicles throughout their service lives. Some of them may have been renumbered more than once. Thus the numbers listed may not have been carried on the full range of coaches at the same time but were carried by the coaches specified for at least a proportion of their time in the valley. The authors would welcome any further information or comments from readers on coach numbering.

Coach numbers

GVT Number	Coach Type	Coach Series
1	Clerestory Saloon	1891[a]
2	Small Open End Balcony Coach	1891[a]
3	Closed Centre Door Coach	1891-2[d]
4	Closed Third Class Coach	1892[a]
5	Open Third Class Coach	1892[d]
6	Closed Third Class Coach	1901[a,c]
7	Closed Third Class Coach	1901[a,c]
8	Closed Third Class Coach	1892[a]
9	Open Third Class Coach	1892[a]
10	Open Third Class Coach	1892[d]
11	Open Third Class Coach	1892[d]
12	Open Third Class Coach	1892[d]
13	Open Third Class Coach	1892[d]
14	Closed First Class Coach	1892[a,b]

[a] based on photographic evidence; [b] now preserved on Talyllyn Railway; [c] one of these two coaches preserved on Tallylyn Railway as first class no. 15; [d] based on circumstantial evidence only.

Modelling the Glyn Valley Tramway Coaches

For those who wish to model some examples of the coaching stock without the necessity for scratch building, kits may be obtained from a number of suppliers. These are outlined in the following directory.

009 scale for 4mm gauge

Langley Miniature Models, 166 Three Bridges Road, Crawley, Sussex, RH10 1LE. Langley offer a selection of kits in etched brass including the First Class Carriage, the 1892 Third Class Coach and the Open Third Class Coach in its final form.

Parkside Dundas, Millie Street, Kirkcaldy, Fife, KY1 2NL. Plastic kits are available for the Centre Door Saloon, the 1892 Third Class Coach and the Open Third Class Coach in its final form. The models provide good panel detail but are a little longer than scaled originals.

Artistry in Brass, address as Parkside Dundas. An etched brass model kit is available for the End Balcony Open Saloon in its original form with side curtains.

0-16.5mm Scale for 7mm gauge

Langley Miniature Models. Etched brass kits are offered for the First Class Carriage, the 1892 Closed Third Class Coach and the Open Third Class Coach in its final form.

Peco, Pritchard Patent Produce Co. Ltd., Beer, Seaton, Devon, EX12 3NA. Peco produce a plastic coach kit which conforms to Glyn Valley Tramway style and dimensions in all respects except for the use of freelance sides. The company also makes a separate plastic underframe kit which provides an accurate basis for the addition of a scratchbuilt body.

Wrightlines, Keith Butler, 36 Field Barn Drive, Weymouth, Dorset, DT4 OED. Keith Butler offers several coach kits with etched brass sides and ends, plastic roof and white metal chassis. The models include the Centre Door Saloon, First Class Carriage, 1892 Third Class Coach and Open Third Class Coach in final form. A white metal lamp top is available separately.

16mm Scale for 32mm gauge or 45mm gauge

Tenmille, 18 Thornley Road, Capel St Mary, Ipswich, Suffolk, IP9 2LQ. The strong, robust, wooden kits are designed for use with live steam garden railways and some scale details have been sacrificed to this requirement. The Centre Door Saloon, a Closed Third Class and the Open Third Class Coaches are available. Nylon axleboxes are also supplied.

J D Models, 10 Rogate Road, Luton, Bedfordshire, LU2 8HR. A full range of partly-completed models are available in wood and plastic allowing the modeller to carry out finishing and detailing. Again, the coaches are particularly suitable for the garden railway, six models are available, Clerestory Saloon, Centre Door Saloon, a Third Class Closed Carriage, the Open Third Class Coach in final form, the First Class Carriage and the End Balcony Open Saloon.